USING DATA SHARING TO IMPROVE COORDINATION IN PEACEBUILDING

Report of a Workshop by the
National Academy of Engineering and United States Institute of Peace
Roundtable on Technology, Science, and Peacebuilding

Andrew Robertson and Steve Olson, *Rapporteurs*

NATIONAL ACADEMY OF ENGINEERING
OF THE NATIONAL ACADEMIES

UNITED STATES INSTITUTE OF PEACE

THE NATIONAL ACADEMIES PRESS
Washington, D.C.
www.nap.edu

THE NATIONAL ACADEMIES PRESS   500 Fifth Street, NW   Washington, DC 20001

NOTICE: This publication has been reviewed according to procedures approved by the National Academy of Engineering report review process. Publication of signed work signifies that it is judged a competent and useful contribution worthy of public consideration, but it does not imply endorsement of conclusions or recommendations by the National Academy of Engineering. The interpretations and conclusions in such publications are those of the authors and do not purport to reflect the views of the council, officers, or staff of the National Academy of Engineering.

The Roundtable on Technology, Science, and Peacebuilding, the sponsor of the workshop on which this report is based, is supported by funding from the U.S. Department of Defense (JDDM-3663-1), Qualcomm, National Science Foundation (ENG-1136841), U.S. Department of Agriculture (59-0790-2-058), U.S. Department of State, and CRDF Global. Any opinions, findings, or conclusions expressed in this publication are those of the workshop participants.

International Standard Book Number 13:   978-0-309-26513-3
International Standard Book Number 10:   0-309-26513-4

Copies of this report are available from the National Academies Press, 500 Fifth Street NW, Keck 360, Washington, DC 20001; (888) 624-8373; online at www.nap.edu.

For more information about the National Academy of Engineering, visit the NAE home page at www.nae.edu.

Copyright 2012 by the National Academy of Sciences. All rights reserved.

Printed in the United States of America

THE NATIONAL ACADEMIES
Advisers to the Nation on Science, Engineering, and Medicine

The **National Academy of Sciences** is a private, nonprofit, self-perpetuating society of distinguished scholars engaged in scientific and engineering research, dedicated to the furtherance of science and technology and to their use for the general welfare. Upon the authority of the charter granted to it by the Congress in 1863, the Academy has a mandate that requires it to advise the federal government on scientific and technical matters. Dr. Ralph J. Cicerone is president of the National Academy of Sciences.

The **National Academy of Engineering** was established in 1964, under the charter of the National Academy of Sciences, as a parallel organization of outstanding engineers. It is autonomous in its administration and in the selection of its members, sharing with the National Academy of Sciences the responsibility for advising the federal government. The National Academy of Engineering also sponsors engineering programs aimed at meeting national needs, encourages education and research, and recognizes the superior achievements of engineers. Dr. Charles M. Vest is president of the National Academy of Engineering.

The **Institute of Medicine** was established in 1970 by the National Academy of Sciences to secure the services of eminent members of appropriate professions in the examination of policy matters pertaining to the health of the public. The Institute acts under the responsibility given to the National Academy of Sciences by its congressional charter to be an adviser to the federal government and, upon its own initiative, to identify issues of medical care, research, and education. Dr. Harvey V. Fineberg is president of the Institute of Medicine.

The **National Research Council** was organized by the National Academy of Sciences in 1916 to associate the broad community of science and technology with the Academy's purposes of furthering knowledge and advising the federal government. Functioning in accordance with general policies determined by the Academy, the Council has become the principal operating agency of both the National Academy of Sciences and the National Academy of Engineering in providing services to the government, the public, and the scientific and engineering communities. The Council is administered jointly by both Academies and the Institute of Medicine. Dr. Ralph J. Cicerone and Dr. Charles M. Vest are chair and vice chair, respectively, of the National Research Council.

www.national-academies.org

The United States Institute of Peace is the global conflict management center for the United States. Created by Congress in 1984 to be independent and nonpartisan, the Institute works to prevent, mitigate, and resolve international conflict through nonviolent means. USIP operates in the world's most challenging conflict zones, and it leads in professional conflict management and peacebuilding by applying innovative tools, convening experts and stakeholders, supporting policymakers, and providing public education. The Institute translates its on-the-ground experience into knowledge, skills, and resources for policymakers, the US military, government and civilian leaders, nongovernmental organizations, practitioners, and citizens both here and abroad.

The Institute's permanent headquarters and conference center are located at the northwest corner of the National Mall in Washington, DC. The facility also houses the Academy for International Conflict Management and Peacebuilding and the Global Peacebuilding Center.

www.usip.org

WORKSHOP STEERING COMMITTEE

Melanie Greenberg (*Cochair*), President and CEO, Alliance for Peacebuilding
Elmer Roman (*Cochair*), Oversight Executive, Office of the Secretary of Defense, US Department of Defense
Gregor Bailar, Chief Information Officer (ret.), Capital One Financial Corporation
Mark Hainsey, Project Leader, US Army Corps of Engineers
Chip Hauss, Director, Alliance for Peacebuilding
Suzanne Kindervatter, Vice President, InterAction
Stephen Lowe, Geospatial Information Officer, US Department of Agriculture
Phuong Pham, Research Scientist, Harvard Humanitarian Initiative
Michael Shipler, Senior Program Advisor, Search for Common Ground

Staff

Geneve Bergeron, Research Assistant, US Institute of Peace
Sheldon Himelfarb, Director, Center of Innovation for Science, Technology, and Peacebuilding, US Institute of Peace
Greg Pearson, Senior Program Officer, National Academy of Engineering
Proctor P. Reid, Director, NAE Program Office
Andrew Robertson, Senior Program Officer, US Institute of Peace
Frederick S. Tipson, Special Advisor, Center of Innovation for Science, Technology, and Peacebuilding, US Institute of Peace

Acknowledgments

This summary has been reviewed in draft form by individuals chosen for their diverse perspectives and technical expertise, in accordance with procedures approved by the National Academies. The purpose of the independent review is to provide candid and critical comments to assist the NAE in making its published report as sound as possible and to ensure that the report meets institutional standards for objectivity, evidence, and responsiveness to the study charge. The review comments and draft manuscript remain confidential to protect the integrity of the deliberative process. We wish to thank the following individuals for their review of this report:

Gregor Bailar, Capital One Financial Corporation (ret.)
Chip Hauss, Alliance for Peacebuilding
David Kamien, Mind Alliance
Phuong Pham, Harvard Humanitarian Initiative
Shannon Scribner, Oxfam America
Michael Shipler, Search for Common Ground

Although the reviewers listed above provided many constructive comments and suggestions, they were not asked to endorse the views expressed in the report, nor did they see the final draft of the report before its release. The review of this report was overseen by Lawrence T. Papay, CEO and Principal,

PQR, LLC. Appointed by NAE, he was responsible for making certain that an independent examination of this report was carried out in accordance with institutional procedures and that all review comments were carefully considered. Responsibility for the final content of this report rests entirely with the authors and NAE.

Contents

1 INTRODUCTION, OVERVIEW, AND THEMES OF THE WORKSHOP 1
   Harnessing Information for a Shared Vision, 3
   Organization and Themes of the Workshop, 4

2 DEFINING THE CHALLENGES OF COORDINATION 7
   A Clash of Cultures, 7
   Data Sharing in Context, 8
   Coordination in Peacebuilding, 10
   Discussion, 11

3 OVERCOMING CHALLENGES TO SHARING INFORMATION 15
   Civilian-Military Guidelines for Interactions, 16
   Civilian-Military Guidelines for Sharing Project Information, 19
   Discussion, 21

4 USING DATA FOR IMPACT 25
   Data Integration and Visualization, 25
   Assessing Value in Data for Development Research, 28
   Understanding Frames of Reference, 29
   Discussion, 31

5	DEMONSTRATING DATA SHARING: THE UNITY SYSTEM	35
	The UNITY System, 36	
	Discussion, 37	

Appendixes

A	Workshop Agenda	41
B	Attendees	45

1

Introduction, Overview, and Themes of the Workshop

On May 23, 2012, the Roundtable on Technology, Science, and Peacebuilding convened a workshop at the United States Institute of Peace (USIP) to investigate data sharing as a means of improving coordination among US government and nongovernment stakeholders involved in peacebuilding and conflict management activities. The following question was the focus of the workshop.

> What needs must a data-sharing system address to create more effective coordination in conflict zones and to promote the participation of federal agencies and nonfederal organizations in peacebuilding?

In addition, the workshop served as a means to obtain feedback on the UNITY system, a data-sharing platform developed by the Department of Defense (DOD) and the United States Agency for International Development (USAID).

The Roundtable was established in 2011 as a partnership between USIP and the National Academy of Engineering (NAE) to make a measurable and positive impact on conflict management, peacebuilding, and security capabilities by bringing together leaders from the technical and peacebuilding communities. Its members are senior executives and experts from leading governmental organizations, universities, corporations, and nongovernmental organizations. Its principal goals are:

1. To accelerate the application of science and technology to the process of peacebuilding and stabilization;
2. To promote systematic, high-level communication between peacebuilding and technical organizations on the problems faced and the technical capabilities required for successful peacebuilding; and
3. To collaborate in applying new science and technology to the most pressing challenges faced by local and international peacebuilders working in conflict zones.

The Roundtable is strongly committed to action-oriented projects, and the long-term goal of each is to demonstrate viability with a successful field trial. The Roundtable has selected a portfolio of high-impact peacebuilding problems on which to focus its efforts:

1. Adapting agricultural extension services to peacebuilding,
2. Using data sharing to improve coordination in peacebuilding,
3. Sensing emerging conflicts, and
4. Harnessing systems methods for delivery of peacebuilding services.

Four steering committees comprised of Roundtable members and other experts developed action plans for each activity area that included workshops intended to assemble experts from across the peacebuilding and technical communities. The workshop held on May 23, 2012, was the second in a series that will address these four topics.

The workshop engaged two types of participant. From the world of conflict management, it included policymakers, planners, and people working in conflict environments. On the technical side, participants included engineers, IT specialists, and analysts (such as econometricians) responsible for the design, use, and maintenance of information technology systems in development and conflict environments. Because the focus of the meeting was on data sharing to improve coordination among federal government and nongovernment stakeholders, the participants came largely from US-based government, NGO, corporate, and academic organizations. This summary should be of interest to a similar audience.

The intent of this summary is to provide an overview of the topics and themes discussed during the workshop. Following further consultation with potential government and nongovernment users of such technologies, USIP will publish a Special Report building on insights from the workshop and identifying the particular capabilities required in a next-generation data-sharing platform to improve coordination in peacebuilding.

HARNESSING INFORMATION FOR A SHARED VISION

According to Melanie Greenberg, president and CEO of the Alliance for Peacebuilding and cochair of the workshop, norms around the creation and use of information sharing[1] are changing rapidly. Significantly more data are being generated and communicated, and these data can be analyzed much more quickly and the results disseminated much more broadly than in the past. For example, data generated in the course of our daily lives, such as information shared on social media sites, can be gathered and parsed to shed light on broad societal developments. Greenberg observed that these new technological capabilities to produce, analyze, and disseminate data are generating moral, ethical, and cultural challenges for producers and users alike. (Chapter 2 presents an analysis of these challenges.)

Additional challenges arise because different organizations have different ways of gathering, analyzing, disseminating, and storing data, said the workshop's other cochair, Elmer Roman, who is oversight executive in the Office of the Secretary of Defense. Even if organizations want to cooperate, cultural differences may erect barriers to doing so.

As an example of an organization that has been working to overcome these barriers, Greenberg cited the Alliance for Peacebuilding, which she heads. The Alliance is a platform for 80 organizations working on a very broad range of peacebuilding issues. To build a sustainable peace, these organizations need to cooperate by sharing information, whether they are involved in peacebuilding, defense, food security, health, science, development, democracy building, civil society building, or some other activity. In fragile and chaotic conflict environments, success requires an "inclusive vision" of what peacebuilding needs to accomplish.

Both Roman and Greenberg emphasized that information sharing can be formal or informal, within an organization or among organizations, and within a sector or across sectors. A common element of successful data sharing is a collective vision among the entities involved and some established common objectives so that they can function as a network of distinct but

---

[1] The Roundtable uses information sharing and data sharing interchangeably.

interconnected agents. "We need to understand where everybody is headed and what's being done," said Greenberg.

ORGANIZATION AND THEMES OF THE WORKSHOP

The workshop was divided into four sessions, with speaker presentations followed by an extended discussion period. In the first session, three speakers identified key challenges in data sharing (Chapter 2). The second session (Chapter 3) featured two speakers who explored ways to overcome these challenges. In the third session, three speakers described specific examples of data-sharing systems (Chapter 4). The final session (Chapter 5) provided an in-depth examination of the UNITY system developed by the Department of Defense (DOD) and the US Agency for International Development (USAID)

Four preliminary themes emerged from the workshop discussions. While these themes will need to be explored more fully before we can draw firm conclusions, they present a reasonable overview of the key issues surfaced by meeting participants.

1. *Data sharing requires working across a technology-culture divide.*
   Coordination among organizations requires ongoing maintenance of relationships, and these relationships depend on sociocultural factors, not technological factors. Technology can provide a means of facilitating relationships, but cannot by itself create them. Rather, organizations that often have different missions, goals, and perspectives may benefit by finding common ground where shared approaches and objectives are possible.
2. *Information sharing requires building and maintaining trust.*
   This trust is built on both technical and social components. "Can I trust the people with whom I am working?" "Will they protect my interests if I provide them with information?" "Can I trust the information I receive to make important decisions?" The basis for trust differs from a technical and a social perspective. In the technical world, the basis for trust is method. In peacebuilding, the basis for trust is communication. Widespread adoption of technologies to support a process as risky as data sharing may require careful negotiation and discussion to create sufficient trust.
3. *Information sharing requires linking civilian-military policy discussions to technology.*

Among nongovernmental organizations (NGOs), trust is built and maintained through continuous engagement, not simply through the provision of functionality. Adoption of collaborative information-sharing technology has to occur through ongoing interaction between the providers of technology and its users, who in this case are peacebuilders. Thus, the activities of civilian-military working groups that have supported interactions between government and civil society in the past may need to be broadened to encompass technology.

4. *Collaboration software needs to be aligned with user needs.*
Finally, software that does not address the needs of peacebuilders is unlikely to be retained over the long term. Continued dialogue between users and technology providers is essential to gauge needs and adjust functionality as needs evolve.

Participants also discussed barriers to coordination: competition for resources, meeting overload, and the lack of a single leader to drive the process forward. But workshop participants agreed that coordination is integral to peacebuilding efforts, as it provides an environment for solving immediate problems rather than in high-level strategic planning. Evaluation of intermediate measures such as the specific outcomes of coordination can keep the emphasis on problem solving.

Formal guidelines are typically necessary to create an environment in which organizations with different missions can share data. In particular, humanitarian organizations often cannot risk being seen as collaborating with the military in high-risk areas or conflict zones if they are to avoid becoming targets in conflicts. A common interest in sharing data is only the first step. Ongoing civilian-military dialogue is essential for groups to build trust and provide channels to address operational problems. For example, negotiations may be needed to develop policies for when to allow open communication between partners or when to permit sharing of other organizations' information outside the stakeholder group.

Actual experiences with methods of data sharing for peacebuilding can make both the challenges and ways to overcome them more concrete. For example, an especially useful way to summarize data is through the use of maps, particularly those that overlay different kinds of data on a geographical grid. In addition, analyses of data from large-scale surveys in countries affected by mass violence can bridge the gap between peacebuilding as intended by policymakers and its implementation and perceptions on the ground.

Institutionalizing and formalizing relationships among organizations can provide a framework for cooperation. And the cooperation itself, including data sharing, is a means to the end of building partnerships and trust and needs to be continually refreshed and renewed.

2

Defining the Challenges of Coordination

The capacity of technologies to gather, sort, and manage data has undergone rapid advances. Yet coordination among government and nongovernment actors in peacebuilding remains a challenge, largely because of the human factors involved in data sharing. Once shared, data can no longer be controlled, and so the problem is not technical but a matter of trust. That trust is based on the capacity to negotiate shared goals, processes, and values for cooperation. The first session of the workshop discussed the ethical, cultural, and social obstacles faced by peacebuilding organizations in adopting technologies to break down the information silos in which they work.

A CLASH OF CULTURES

Tremendous progress has been achieved over the past 20 years in improving the sharing of information among organizations involved in peacebuilding, said Ambassador Robert Loftis, Interagency Professional in Residence at USIP. But even after a decade of experience in Iraq and Afghanistan, and more than two decades of peacebuilding activities since the fall of the Berlin Wall, challenges of coordination are still prominent.

A major contributor to these challenges, Loftis explained, is that peacebuilding is marked by a clash of cultures. Military, civilian government, and nongovernmental organizations (NGOs) do not have the same immediate goals and timelines. In many cases, they do not even speak the same language. For example, to the military, "coordination" means the ability to direct; to civilian organizations, it means consultation and consensus; and to NGOs, it means avoid contradictory activities or—more positively—sharing information. These differences can be strength, in that different organizations bring different perspectives to a problem, but unity of effort may be an elusive goal.

Some of these challenges can be addressed by institutionalizing and formalizing relationships, Loftis said. For example, a mechanism such as the Civilian Military Relations Working Group in Nonpermissive Environments, which brings together a variety of organizations for discussions every few months, can provide a framework for cooperation. But such mechanisms are means to an end and not ends in themselves. They are typically built on personal relations that do not necessarily carry over as personnel change. Personal relations and trust must therefore be continually refreshed and renewed. A relationship that worked in one context is not guaranteed to work in another, even with the same individuals. And as great as the challenges are in a purely American context, they are multiplied many times over in multinational or multilateral environments.

Data sharing can be useful in building relationships and trust, Loftis concluded. It can start on a small scale and gradually expand as trust and experience build. But assumptions and expectations about the use and distribution of data must be made explicit early, some data will be easier to share than others, systems have to evolve over time to be effective, and data sharing is no substitute for critical thinking and communication.

DATA SHARING IN CONTEXT

To understand the challenges of information sharing, it needs to be seen in the context of the broader structure and experience of civilian-military relations and civilian management of humanitarian, development, and peacebuilding activities, said Randy Tift, senior policy advisor in the World Vision US International Programs Group. For NGOs, information sharing occurs along a spectrum of involvement, from information gathering and needs assessment to the use of information for the delivery of aid or services. As an example of the latter, Tift cited the earthquake in Haiti: NGOs responded to demands from the UN, the United States, and the host government for information and coordination, resulting in a much greater degree of donor coordination.

Tift focused on two areas: the security of NGO staff and the people they serve, and community acceptance of NGOs. He noted that information sharing can be a powerful determinant, for better or worse, of security. For that reason, as spelled out in World Vision's policies on information sharing and liaison arrangements, the organization actively seeks to sustain open and direct (or indirect) dialogue with militaries and other armed groups in all circumstances, always with the objective of protecting civilians and enhancing mutual understanding of roles and mandates.

World Vision also establishes a mechanism for liaison with military actors in situations where it shares operational space with such groups. Liaison may take place through a coalition of NGOs, established lines of communication maintained by the UN, or direct communication when appropriate. Liaisons need to be transparent to all stakeholders and maintain a clear distinction between armed actors and NGO workers.

In humanitarian operating environments, World Vision and military or police personnel need to maintain a mutual understanding of objectives, roles, activities, and principles, said Tift. World Vision seeks to engage in ongoing dialogue with the military and police, with a view to promoting adherence to international humanitarian law and other human rights instruments and to increasing the military's understanding of the roles of humanitarian organizations.

World Vision recognizes that in some cases military and international police actors are in a unique position to provide data about specific humanitarian needs. In cases of extremely vulnerable populations for which data are lacking, World Vision and other organizations have sought out this information. However, the information has to be triangulated with that from other sources to confirm its reliability. The data have to appear, without reasonable doubt, to address a humanitarian imperative. Military or armed police contingents should not be able to gain legitimacy simply because they have a relationship with an international humanitarian organization.

The same considerations apply to data flows in the opposite direction, Tift said. To maintain credibility with local organizations and individuals, NGOs cannot appear to be gathering intelligence for the US government or to be functioning as agents of government security operations. NGOs depend on being perceived as impartial, independent humanitarian organizations. Moreover, to be effective, it is critical that NGOs be viewed by host country leaders as trustworthy, above reproach, and committed to addressing underlying causes of poverty and injustice.

Coordination with the US military or other armed actors, if performed on the basis of NGO independence and impartiality, does not necessarily compromise an NGO's acceptance by local communities, Tift said. Indeed, it may help to ensure the effective delivery of aid to victims of poverty and injustice in a complex emergency situation. Enabling NGOs to act independently, even when implementing programs funded by the US government, is not only necessary but makes the achievement of US strategic objectives much more likely. NGO independence does not bind NGOs operationally to security imperatives but rather strengthens US security by addressing root causes of insecurity, according to Tift.

Humanitarian organizations can learn much from each other, Tift concluded. As part of this learning, commitment to better coordination would bring greater unity to NGO initiatives.

COORDINATION IN PEACEBUILDING

Even with coordination among organizations, peacebuilding can be effective or ineffective depending on the context and on how coordination is approached, explained Susanna Campbell, Visiting Scholar at the Saltzman Institute for War and Peace Studies at Columbia University.

Peacebuilding is difficult, she said, and the determinants of peace—even the definition of peace—are not fully understood. Therefore, simply combining the hypothesized elements of peace through coordination will not ensure success. Coordination needs to be aimed at specific problems that stand in the way of peace.

Campbell described several ways in which coordination can lead to ineffective peacebuilding. It can decrease flexibility and the capacity to adapt strategies and approaches, especially if organizations lose touch with the context in which they are working. It can focus attention on other international actors rather than peacebuilding problems that must be solved. Where data are not available on the effects of peacebuilding activities, coordination can lead to uninformed decisions. Coordination efforts can compete for funding with other activities, including those more directly focused on peacebuilding, and can overload organizations' already full agendas.

However, Campbell continued, coordination also can contribute to effective peacebuilding. It can direct attention to peacebuilding efforts that emerge from the bottom up rather than from top-down directives, especially to the extent that such efforts aim to solve immediate problems. It can serve as a forum for stakeholder dialogue and break down cultural barriers between organizations. And coordination can allow organizations to work in a complementary fashion to solve the problems at hand.

Campbell emphasized, as did Loftis, that coordination is a tool, not the end goal of effective peacebuilding. It therefore needs to be judged by intermediate measures such as the focus of coordination, who is involved, and what actions result. At the most basic level, coordination can prevent duplication of activities. At a more ambitious level, it can enable joint action. Assessments of the impact of coordination need to keep these different objectives in mind, and should also take stock of the effects of peacebuilding on the people who are the subject of those efforts.

As one element of coordination, data sharing can prevent duplication and increase the participation of stakeholders who are not traditionally included. Data sharing also can support informed discussions by providing information about outcomes, thus enabling programs to evolve based on their impacts. In this way, organizations can see how a situation is changing and adjust their actions accordingly.

DISCUSSION

During the discussion period, the three speakers and other workshop participants explored the varied challenges to data sharing and coordination in peacebuilding. Tift reiterated that some forms of coordination can actually lead to conflict, such as when they create a perception of alliance with a belligerent party. Some forms of data sharing or other kinds of collaboration with the military or other government agencies can be appropriate, Tift acknowledged, but not if they undermine humanitarian objectives. As an example, he cited a case in Afghanistan where World Vision was asked, as part of a US government grant, to retarget the beneficiaries of its aid to serve counterinsurgency objectives. World Vision refused to do so, as did several other NGOs, which led to a dialogue that ultimately changed the policy.

Campbell observed that even when data sharing is beneficial, it can be difficult to do systematically. For example, NGOs are more effective if they share data among themselves, and they can do so through such means as the UN Office for the Coordination of Humanitarian Affairs (OCHA). But once peacebuilding starts, OCHA tends to recede into the background, and no single organization is charged with collecting and disseminating information. Even when data are shared, she continued, the level of detail often is not sufficient to achieve effective coordination. Furthermore, data on impacts or outcomes are exceedingly scarce.

To deal with these problems, Campbell suggested that much more analysis be done of the institutions that are the focus of change. Who are the key players, what needs to happen, and what systems need to change? Once this analysis has been done, incentives and motivations need to be created to foster change. Then the effectiveness of these interventions needs to be measured. "Was this the right approach? If not, what approach might work better?" Military culture is more amenable to evaluation than is the development community, said Campbell, but in either case representative stakeholder dialogue focused on outcomes can yield the information needed to adapt and learn.

Tift identified another missing ingredient: effective policy dialogue among organizations, especially between NGOs and civilian agencies in the US government. He cited several cases in which policies with a major effect on NGOs were disseminated by US government agencies without consultation with the groups most affected by those policies.

Loftis pointed to some of the deeper problems with measuring impacts. Individuals and organizations want to have an impact and often interpret change as a direct consequence of their activities. However, they cannot know all the factors that came into play. In complex and quickly changing environments, it can be very difficult to determine causality—that a particular action had a particular outcome. Establishing a track record over time can point toward effective action, but evaluation remains a difficult task.

Kevin Brownawell, interagency professional in residence at USIP, observed that organizations often have different ideas of what data to collect. Even different agencies in the US government focus on different aspects of conflict situations and may request different kinds of information from the NGOs with which they work. Conflicts can be avoided by planning for data collection at the beginning.

Gregor Bailar, retired chief information officer for Capital One Financial Corporation, commented on the similarities between the problems discussed in the workshop and the challenges facing large and innovative companies. They, too, encounter problems caused by rigid strategies, lack of coordination, and false confidence in strategic planning, and they, too, benefit from stakeholder dialogue, bottom-up coordination, and organizational integration. Agile approaches to planning and problem solving work well in both the private and public sectors, he said.

Campbell noted that all organizations have difficulty with behavior change, in part because they learn through re-established routines—"Organizations learn what they already know." Therefore, one way to change what institutions learn is to change what they know. If the military wants to engage in peacebuilding, it needs to bring in people who are familiar with peacebuilding activities and make peacebuilding a priority in the military culture. In this way organizations can adapt based on what they learn, though this typically works best in smaller and more agile organizations. In larger organizations, change can occur in pockets of the organization that encourage adaptation and learning.

Finally, Sheldon Himelfarb, director of the Center of Innovation for Science, Technology, and Peacebuilding at USIP, asked about the benefits of transparency. In the Facebook era, an emerging paradigm may be to default to the release of information so it can be used by others. Campbell responded that transparency would be "a huge step forward" but does not necessarily address the full range of problems.

Loftis also cited the problem of too much information. Part of the challenge, he said, is to filter meaningful from meaningless information and to synthesize information in ways that are useful.

3

Overcoming Challenges to Sharing Information

Notwithstanding the challenges associated with data sharing and coordination, the federal government, nongovernmental organizations (NGOs), and other organizations have developed and improved processes for managing their interactions. They have created guidelines that enable cooperation while protecting groups' independence and security.[1] They have instituted processes that enable cooperative planning while maintaining executional autonomy. And they have established agile and bottom-up forums that build the trust necessary for peacebuilding.

In the second session of the workshop, two speakers described how the challenges to sharing information for peacebuilding are being overcome. In the US wars in Iraq and Afghanistan, the issue of cooperation between government and civil society groups has focused on civilian-military interaction. For some NGO organizations, however, any interaction with government may present a problem.

---

[1] See, for example: http://www.usip.org/files/resources/guidelines_pamphlet.pdf and http://ochaonline.un.org/afghanistan/CivilMilitaryCoordination/tabid/5356/language/en-US/Default.aspx.

CIVILIAN-MILITARY GUIDELINES FOR INTERACTIONS

One goal of data sharing is to create a "whole of society" approach in which civil society produces a government that is citizen oriented and not just elite oriented, said Lisa Schirch, founding director of 3P Human Security, a collaboration of the Center for Justice and Peacebuilding at Eastern Mennonite University, the Alliance for Peacebuilding, the Kroc Institute at Notre Dame University, and the Global Partnership for the Prevention of Armed Conflict. In such an approach, a citizen-oriented state, the private sector, and civil society organizations cooperate to promote good governance, development, security, and respect for human rights, and civil society partners with, complements, and supplements government in running programs. Without an active civil society, an elite-oriented state and private business sector can result in instability, corruption, and diminished human rights, as is the case for many countries around the world, Schirch observed.

Civil society does not consist just of NGOs. It includes universities, religious organizations, media, professional associations, trade unions, traditional and tribal organizations, and many other entities that seek to improve quality of life. All of these institutions hold government to account, said Schirch. When civil society does not exist or is quashed by the state, government is no longer accountable. Thus, the members of organizations in this sector serve the public in ways comparable to public sector employees. "They often have just as many credentials and take just as many risks as people in the military," said Schirch.

Yet military personnel often do not understand and sometimes do not even like NGOs, as illustrated by comments quoted by Schirch: "NGOs clog up my battle space." "They are in the way." "NGOs will only call when they need rescuing." "NGOs don't want to be seen with us in uniform." "They don't have the courage to show who their friends are." She noted that stereotypes extend in the opposite direction as well, and that both sets of stereotypes are damaging.

Fostering Dialogue

To foster collaboration, Schirch's organization supports civilian-military dialogue. In particular, she has been connecting military and civil society organizations in Afghanistan, including an Afghan NGO that does mediation and conflict resolution as part of disarmament, demobilization, and reintegration. Staff of the NGO talk with insurgents to determine grievances and how they can be mediated so that the insurgents can reintegrate into society.

Another local NGO has worked on relationships between the police and the communities in which they work. Police personnel meet with community representatives in a facilitated mediation, an approach that has worked so well that the UN Development Program has partnered with the NGO to make the program national.

These models have been very successful, but they have required that Schirch and her colleagues go to the local International Security Assistance Force (ISAF) base every day to share information about the collaboration efforts, as the ISAF personnel were unwilling to make the trip. Schirch and her staff walked a seven-block lane where taxis refused to go. Though she has felt safe in most of Afghanistan, this is "the most dangerous road in Kabul," she said, showing a photograph of the street. "Along that road, I am a free-range target for anybody who wants to kill somebody who's collaborating with ISAF."

In this case, both sides want to share information, but there is no institution, location, or mechanism for them to do so easily and safely. Recently, the UN Assistance Mission in Afghanistan opened bases around the country to facilitate coordination within sectors, such as among groups working on rule of law. "It provides a better place, but that's been a more recent innovation," said Schirch.

Schirch noted that she has lost many colleagues in Afghanistan to violence, showing a photograph of the British cemetery in Kabul where a number of her colleagues are buried. "It's a good month if I don't lose a colleague in Afghanistan or Pakistan." Furthermore, attacks against NGO personnel and other representatives of civil society are increasing, she said, moderated in the last few years only by a withdrawal of NGO personnel from these areas.[2]

What in part is driving these attacks on aid workers, Schirch explained, is a shift by military actors and governments to using development activities as a means of enhancing security and stability following a conflict. Realizing short-term security objectives, however, has an unintended consequence of politicizing the activities of the civil society organizations partnering in development. Civil society organizations try to maintain their workers' safety by not taking sides in conflict and working to relieve all social suffering. Participation in development work with political ends involves aid workers in the underlying disputes and has led to higher levels of violence directed at these individuals.

---

[2]The Afghanistan NGO Safety Office tracks violence against NGO workers at www.afgnso.org.

This difference in objectives is what makes cooperation and data sharing between government and civil society groups difficult. Policy guidance from USAID, State, and the Department of Defense for how government agencies should work with civil society group tends to show little understanding of these divergent goals, she said. Instead, many representatives of government agencies and military forces are eager to use these organizations in the final phase of the "shape, clear, hold, and build" approach to counterinsurgency. They want to use these organizations to work in communities to spur development and thereby fend off insurgency.

NGOs do not want to be the implementers of a security strategy for government, said Schirch. They want conflict assessment and planning to reflect the realities of what they see on the ground. They are willing to provide insights into local dynamics that either help or hinder the protection of human security, but they do not want to be involved in implementing a counterinsurgency strategy.

In general, the goal of civil society organizations is to protect human security, whereas the goal of government agencies and the military is to advance national security interests. Sometimes these goals overlap, but in other cases they conflict, and conflicts produce tension between governments, military forces, and civil society organizations.

Data sharing is much more likely in situations and contexts where the missions of civil society organizations overlap with those of the military and goverment. When missions are in conflict, data sharing is more difficult, said Schirch.

Guidelines for Cooperation

In 2005 the heads of major US humanitarian organizations and US civilian and military leaders met at USIP to initiate a dialogue that led to the Guidelines for Relations Between US Armed Forces and Non-Governmental Humanitarian Organizations in Hostile or Potentially Hostile Environments. These guidelines make it safer for NGOs to do their work without being seen as political actors with short-term security goals. Schirch cited several key elements of the guidelines:

- Visits by US armed forces personnel to NGO sites should be by prior arrangement.

- US armed forces should give NGOs the option of meeting with US armed forces personnel outside military installations for information exchanges.
- US armed forces should not describe NGOs as "force multipliers" or "partners" of the military.
- Vehicles and clothing should distinguish NGOs from the military.

A drawback of the guidelines is that they do not necessarily explain why particular arrangements are important, said Schirch. Also, neither the guidelines nor the rationale for them are routinely taught in military academies or other settings.

Continued civilian-military dialogue will be essential to enable data sharing in peacebuilding, Schirch concluded. The participants in these dialogues—from government, the military, and from civil society—need to develop a better understanding of how their goals and the constraints under which they operate may differ. Schirch said the amount of bad blood she sees between government and civil society groups now is tremendous. Only by a shared trust-building process, in which both sides see and respect the reasonableness of the other's goals, can we build relationships that support data sharing. Put another way, Schirch asserted, in the current situation in Afghanistan, any technology-based data-sharing approach would be very difficult given the lack of trust. Only by committing to dialogue that allows participants to understand each other can we realize a whole-of-society approach. By communicating goals, plans, and worldviews, government, military, and NGO stakeholders could develop a shared understanding that would better promote unity of effort.

CIVILIAN-MILITARY GUIDELINES FOR SHARING PROJECT INFORMATION

Marcia Hartwell, Visiting Scholar at USIP, drew on her experiences in Iraq to address the development of civilian-military guidelines for sharing project information. Information sharing is a hot-button topic for everyone and one of the most sensitive topics in a conflict area, she said.

Hartwell explained that access to project information can be either open or controlled. In addition, each organization's internal use of unclassified information, which can have many layers of sensitivity and confidentiality, is a consideration in sharing that information. Because information is often power, its actual versus intended use can be an important factor to consider.

The use of information can have positive or negative consequences. Will the military use information provided by civilian organizations for targeting purposes? Will civilian organizations withhold information from the military that could turn military forces into targets? There are sensitivities on both sides, said Hartwell.

One way to explore and understand these sensitivities is to establish guidelines for a vetting system that identifies and monitors potentially sensitive information. For example, all data providers and users of project data could receive online conflict awareness training. Will the public or private posting of project details potentially endanger anyone on the ground? What precautions can be taken to avoid this? What are inflammatory issues that could stoke ethnic or sectarian divisions? Even issues as seemingly innocuous as access to water can inflame an entire region, Hartwell said, as can access to infrastructure—if, for example, economic assistance appears to favor one group over another. "Understanding potential flash points, and how those flash points move around and evolve…is a big issue."

Hartwell also spoke about managing expectations. Short- and long-term goals for information sharing can build sustainable information-sharing networks, but they need to reflect both similarities and differences in civilian-military timelines, capacities, and missions, she said. Sustainability for the military is often short term, whereas NGOs tend to look at issues in a more open and extended context.

Hartwell advocated the establishment of a civilian-military data-sharing working group with several goals:

- View data sharing as a long-term process of building trust between civilian and military organizations.
- Assist in defining and negotiating virtual and real space during interventions.
- Clarify how this information could contribute to decision making in future civilian-military interventions.

Establishing such a working group will be particularly important as the military withdraws from Afghanistan and begins to turn certain operations over to civil organizations, she said. Discussions can help determine whether such handoffs are appropriate or advisable. In addition, predicting a future of more small-scale conflicts where the civilian-military nexus will grow, Hartwell concluded that "it is incredibly important that we learn to work together in a real and honest way."

DISCUSSION

Suzanne Kindervatter, head of the Strategic Impact Team for InterAction, described a "wall of honor" at the organization's headquarters building, with the addition each year of the names of more NGO staff members killed in conflict situations. She said that the civilian-military guidelines described by Schirch were "a major breakthrough" and that the process of developing them recognized the differences in culture and missions and then built a set of terms of references for different groups to work together. These guidelines need to be publicized, she said, for them to have the effect that they should.

Kindervatter acknowledged that even within the world of NGOs, much more work is needed to promote data sharing. These organizations have different data systems and terminologies; for example, what one NGO calls a project another might call a program. In addition, trust issues arise even among NGOs. Greater transparency and building on past successes can help overcome these barriers.

Linton Wells, director of the Center for Technology and National Security Policy at the National Defense University, said that his office has been looking at the difference between unclassified and nonclassified information. Unclassified information has been reviewed by a classification authority and determined to be unclassified. Nonclassified information has not been reviewed but may be extremely valuable. As an example, he cited social media exchanges between a hospital, an ambulance, and an NGO about an event involving injuries. Use of this easily accessible nonclassified information may undermine trust if the information appears to have come from sensitive sources. Thus, information is not necessarily good or bad, he said, but can be used by different people in different contexts for different purposes.

Schirch observed that once information is available its use generally cannot be controlled. Also, people in the countries where NGOs are operating are usually far more sophisticated about the use of information than outside groups assume. They know how to manipulate information networks, use the media, and interpret information. In addition, they have their own information sources that are far more effective than the ones to which outsiders have access. For example, a village or marketplace may suddenly empty of people before outsiders know anything about a threat. A better approach than trying to control information is to address the vulnerabilities for both the civilian sector and the military sector regarding the use of information.

Schirch also pointed out that many NGOs are hesitant to take Western donor money and therefore seek other sources of funding because of safety and security concerns. Many NGO representatives are kidnapped and held for 72 hours, during which the kidnappers are checking the web to see who they are, who they work for, and the source of their money. For this and other reasons, most NGOs do not want to be associated with the military on websites, because they can then become targets. NGOs that are explicit about being part of a counterinsurgency campaign are "hit all the time," said Schirch, and have much greater security concerns than do NGOs that are more careful about their relationship with the US government.

Hartwell reiterated her point about the value of online conflict training, which can help the military understand how violence unfolds across all of society. The lack of such training has been "a real vulnerability in almost all of the strategic and operational initiatives that the US government has tried to implement."

Michael Shipler, Asia director for Search for Common Ground, noted that innovation often emerges from the interaction of groups of people who work on very different things and have very different frames of reference and worldviews. Peacebuilders are searching for transformative innovations that can magnify their influence, which requires interactions encompassing groups that range from the field and operational level to the policy level.

He also identified obstacles that prevent groups such as Search for Common Ground from sharing information with military organizations. Because these groups are committed to doing no harm, when they have information about who has been recruited as a soldier, who has been victimized by an armed group, or who is serving in a group that may be an enemy of the United States, they may not share it even though it could be very useful intelligence.

Shipler reaffirmed that peacebuilding organizations have to remain impartial to be effective, and sharing information with one side but not another may compromise this ability. Groups may lose their comparative advantage as peacebuilding organizations if it were known that they were giving information to one side or the other. By remaining independent, such organizations can have access to many types of information; in fact, one measure of their value is the ability to convince people who have information that the organization will not pass that information on to others. For example, local groups need to trust an organization to tell them the mechanisms through which recruitment to insurgency groups is occurring. "The protection of those sources and of that information is something that builds trust over time," said Shipler. If this trust is violated, people go quiet or stop creating access to information.

In response to a question from Elmer Roman about how to move forward so that the same issues are not being discussed 10 years from now, Hartwell advocated the establishment of an entity with concrete goals that would work on small but specific questions: What information is going to be shared in a particular country, how, and with whom? As specific questions are answered, the scope of the discussions could be broadened.

Frederick Tipson, Jennings Randolph Senior Fellow at USIP, pointed out that the technology in the battle space and the peacebuilding space is very different now than it was just a decade ago, and this change has transformed the challenges of building peace. He advocated local, customized trials of how information can be shared in given settings. Each situation is different, and technology can allow each peacebuilding approach to be customized and localized.

Aaron Chassy, senior technical advisor, Governance and Civil Society, for Catholic Relief Services, observed that power is not balanced between US government forces and civil society organizations. In particular, the law against providing material support or resources to a foreign terrorist organization, even if the intention is to support a group's humanitarian efforts, has very serious implications for organizations like Catholic Relief Services. Such organizations want to bring people back into civil society so that they can choose nonviolent means to resolve their conflicts. "We can't do that if we have US criminal law weighing over our heads that would destroy us as an agency if we went ahead and had that dialogue."

One way to avoid working with groups that want no association with the US government is to work through local partners, Chassy said. But because such partners may have ties with terrorist organizations, it puts a very heavy burden on civil society organizations to do due diligence. "We certainly want to be willing and capable partners with a whole-of-society approach," said Chassy, "but there needs to be a more level playing field and shared and mutual accountability."

Schirch agreed that this is a difficult problem, noting that she teaches at Kabul University and worries that she will be arrested for teaching students who are sympathetic to the Taliban, even though what she is teaching "is absolutely not fueling the insurgency."

Hartwell emphasized the importance of communicating intentions, which is also part of sharing information. If the military knows that an NGO representative is engaged in teaching, that can be enough information to avoid problems.

Julie Montgomery, director of innovation and learning at InterAction, said that to protect staff, her organization often does not indicate their exact locations on publicly available information sources. The question then becomes whether to have public and private databases, where some information is shared only with certain groups.

Susanna Campbell added that the type of data and the operating environment need to be distinguished. In some countries, the US military or other parts of the US government are heavily involved, while in other countries they are not. The risks involved in the exchange of a given type of information change as the degree of violence changes.

Eric Gundersen, president and cofounder of Development Seed, pointed to factors other than security that can influence information sharing. One is interoperability. When information is published in the form of PDFs, for example, it typically needs to be loaded by hand into databases, which can severely slow data sharing. In other cases, key data are missing, are not available, or can be obtained only from sources that cannot be publicly cited.

Gregor Bailar, retired chief information officer of Capital One Financial Corporation, pointed to factors that affect the value of data: timeliness, reliability, completeness, accuracy, insight, and actionability. And the single most important attribute of valuable information is that it comes from a trusted source. Really valuable data are scarce, he said, and relatively few people have access to those data. Also, operational data can be much more valuable than data generated through a research study, because operational data can always be fresh.

Hartwell observed that researchers often need to protect vulnerable sources, which means that researchers need to be trusted for their results to be seen as credible. She added that the turnover of personnel is "a huge issue," not just in civil society organizations but in the military. People can be doing great work, "and then they leave and take everything with them." And new people coming in may not know about work that has been done earlier, which can be tremendously frustrating. One way to avoid such situations is to establish protocols for ongoing information gathering and dissemination that everyone understands so that knowledge of procedures and findings extends beyond current personnel.

Finally, Roman emphasized that NGOs and the military may have different missions, but that NGO personnel are "great American heroes" to the military. "We share a common goal," he said. "In the end, it boils down to human security."

4

Using Data for Impact

Multiple challenges impede data-sharing efforts. Beyond differences in organizational culture and mission, simply establishing the processes that allow collection and sharing of data between organizations can be costly and time consuming. To overcome these challenges, managers of data-sharing systems must consider how the data are to be used and how to get the most impact from the data. Only by improving the impact of shared data can better incentive be created for broad and sustained participation in data sharing. In the third session of the workshop, three speakers provided examples of approaches to improve the impact of data sharing.

DATA INTEGRATION AND VISUALIZATION

New technologies have created the ability to gather, integrate, visualize, and disseminate data in ways that are qualitatively and quantitatively different from what has been possible before. Patrick Vinck, research scientist at the Harvard Humanitarian Initiative, described some of these new capabilities.

New Software

Thanks to new software, data collection and analysis are increasingly characterized by both precision and speed of acquisition. Advances in software also enable rapid progression from data collection to analysis and dissemination, which can allow results to feed back into data collection. It is frequently possible to move from virtually no data to comprehensive data in a very short time. For example, an entire city can be mapped in just a couple of days using street mapping software and volunteers who are motivated to collect, assemble, and present information.

One risk of new methods of data collection is that the amount of data collected can be overwhelming. Data therefore need to be aggregated and summarized. "I say 'summarize' instead of 'simplify,'" said Vinck, because data need to be made more consumable without decreasing their value.

An especially useful way to summarize data is through the use of maps. For example, the LRA Crisis Tracker is a real-time data collection and mapping platform that tracks the atrocities of the Lord's Resistance Army in Africa.[1] Vinck also cited the Satellite Sentinel Project, in which the Harvard Humanitarian Initiative is involved, that seeks to deter atrocities by focusing world attention on threats to civilians.[2] This project uses what a few years ago would have been military-grade satellite data for the purposes of protection and warning.

Another project of the Harvard Humanitarian Initiative is PeacebuildingData.org, which seeks to give a voice to the people involved in peacebuilding and reconstruction processes. It features analyses and data from large-scale surveys in countries affected by mass violence and aims to bridge the gap between peacebuilding as intended by policymakers and its implementation and perception on the ground. Survey takers seek answers to questions such as: What have people experienced? How is the peacebuilding process affecting them? What do they think should be done? The information is collected digitally, which makes it faster to produce and results in better quality. Working in just a few countries, the project has sought to build a baseline of information that can be revisited every few years to gauge changes. It also can single out individual projects to determine whether they have been successful or not. An important application of such efforts is to help determine the extent to which the investments of the international community have led to peacebuilding.

---

[1] See http://lracrisistracker.com.
[2] See www.satsentinel.org.

Data Platforms and Network Architectures

Vinck emphasized the value of letting data speak for themselves rather than having researchers create a narrative. By presenting data through a technology platform, users can interact with the data. Vinck pointed to the PeacebuildingData.org project in Liberia (Figure 4-1), in which an online data set presented using Google Maps allows users to create indicators that are of interest to them. Similarly, for a project in Mindanao, Philippines, users can click on a list of indicators to access and visualize the information they want.

So far, the information in the databases has come from a single source, but Vinck discussed the possibility of layering information from multiple sources onto a single map. Major questions that must be answered for such a system are whether information can be integrated and whether it is useful to do so. In part, he said, the answers depend on the purpose of the project. For example, a project focused on conflict analysis may differ from one focused on communication. Similarly, one project may lend itself to the development of a composite indicator that provides a peacebuilding score, while such an indicator might not be appropriate for a different project.

Ideally, the data presented through interactive platforms would be completely open to users. But data can be expensive and time consuming to collect, and letting go of data can be difficult. Data may also need to be protected

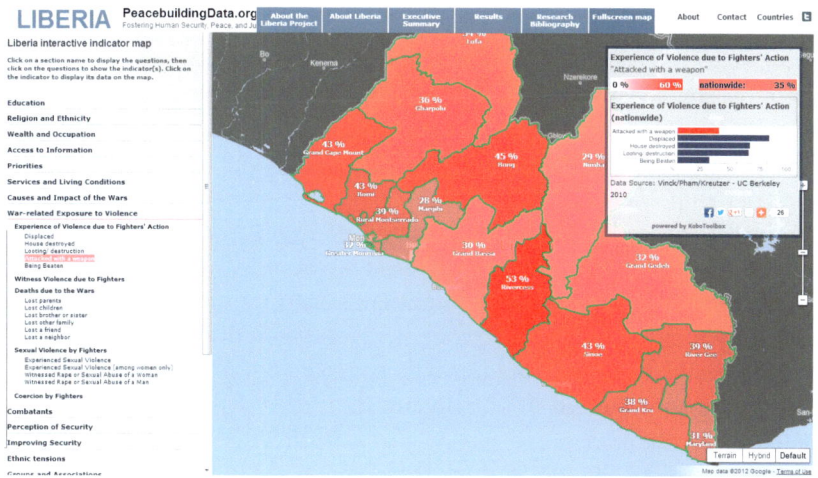

FIGURE 4-1 Survey-Based Conflict Indicators for Liberia SOURCE: PeacebuildingData.org.

if they are from a sensitive source. PeacebuildingData.org has decided to let each organization display its own data rather than collecting the data in a centralized system. That way, each organization retains control over its data and can even choose to withdraw the data. Such an approach has implications for updating data, Vinck acknowledged, as the information displayed can have different time frames and references. The development and use of metadata can help interactive platforms move toward networks of data sharing in which data have different sources but can be directly compared.

Ethical Implications

The collection, analysis, and sharing of data have important ethical implications, Vinck observed. For example, NGOs have various procedures for dealing with sensitive information, but those procedures typically are not shared with the public. How is information vetted? Are human subjects being protected? The lack of transparency makes it difficult to answer these questions for individual organizations. It also makes it difficult for organizations to learn from each other. Widely disseminated guidelines, along with training on how to access and share information, can help organizations deal with issues that arise.

ASSESSING VALUE IN DATA FOR DEVELOPMENT RESEARCH

For data sharing to be effective, the data must have value. Innovations for Policy Action (IPA) is an NGO dedicated to demonstrating the value of data by discovering what works to help the world's poor. It designs and evaluates programs and provides hands-on assistance to bring successful programs to scale. It has more than 200 ongoing projects in about 40 (mostly developing) countries and offices in 14 countries.

Niall Keleher, IPA's director of research methods and training, explained that the organization's long-term mission is not only to identify innovative social programs but to conduct multiple evaluations of programs in order to identify their impact in various contexts and with diverse populations.

IPA begins by identifying not only the intervention to be used but the theory of change behind it. The organization then defines a representative sample for data collection, with particular attention to ensuring that the data accurately capture the populations about which statements are to be made. It does sample size and power calculations, applies valid randomization methods to the population, and develops indicators to accurately measure the value of a program. Computer-assisted interviews provide access to data for prompt quality checks and offer the potential for more timely analysis.

Distinguishing between causality and correlation is a challenge, Keleher acknowledged, but IPA's research design is carefully constructed to elicit "a true and unbiased estimate of the causal relationship between interventions and outcomes."

The Approach in Practice

Keleher described two examples from IPA's portfolio of projects. In northeastern India, the organization measured the success of an NGO seeking to achieve full immunization of children. It found that for an intervention organized around immunization camps where mothers brought their children for immunizations, the full immunization rate jumped from 6 percent for the control group to 18 percent for the group subject to the intervention. Furthermore, when the mothers received a one-kilogram bag of lentils as an incentive, the percentage jumped to 39 percent. "This kind of study is what we aim to produce—something that shows how effective a particular program was."

A study in Malawi looked at the effect on repayment rates of tracking borrowers via fingerprint scanning technology. By having a photograph of a person and a fingerprint in a database, the highest-risk borrowers substantially increased their repayment rates, enabling others to obtain loans whereas before they might have been denied because of risk.

Access to some of the data collected by IPA is limited by confidentiality and intellectual property considerations. But in general IPA seeks to make its data available through publication in scientific journals. This transparency encourages others to try to replicate the evaluation technique, validate the published data, and build on previous results.

UNDERSTANDING FRAMES OF REFERENCE

Stephen Lowe, geospatial information officer in the Office of the Chief Information Officer at the US Department of Agriculture, discussed some of the issues that arise in interagency data sharing. First, he said, factors mentioned in the discussion of data sharing and peacebuilding occur across the federal government. Many government agencies and personnel have different frames of reference—agendas and ways of communicating—yet they face common issues involving data. Where should data sharing start and stop? What are the scope and scale of data sharing needed for a given project? When should data sharing focus on interpretation, and when on discovery? Are the facts available but extremely complex, or are missing facts creating uncertainty? (As Lowe said, "Sometimes we don't know what we don't know.") What is the appropriate tradeoff between data precision and speed of acquisition? Some data are more valuable when they are acquired quickly, as opposed to gathering more precise data over a longer time frame.

Lowe also mentioned a more fundamental difficulty with data sharing: the distinction between policy disagreements and policy controversies. Policy disagreements involve disputes in which the two parties are able to resolve the questions at the heart of the dispute by examining the facts of the situation. Policy controversies are disputes that are immune to resolution by appeal to facts, making them much more intractable. Furthermore, people can focus their attention on different facts or interpret the same facts in different ways, and they have a remarkable ability, when embroiled in a controversy, to dismiss the evidence cited by their antagonists.

Seeing Outside the Frame

People tend to interpret evidence based on the frames of reference they apply, Lowe explained. These frames incorporate beliefs, perceptions, and appreciations that underlie policy positions. They have normative implications that a certain type of solution is acceptable.

To overcome barriers created by different frames of reference, people need to seek agreement on the nature of the problem and the general character and content of a solution, said Lowe. The type of problem to be solved may involve diagnosis, classification, analysis, the detection of anomalies, the configuration or selection of data, monitoring, prediction, design, or planning. Understanding the type of problem leads to better alignment with different types of available solutions.

Lowe further explained that understanding the framing of a problem can create opportunities to operationalize solutions. Some problems lend themselves to customized one-of-a-kind solutions, while others may yield to highly standardized and routine solutions. By moving toward the latter, unit costs can be reduced and efficiencies realized. Understanding the framing correctly can enable the proper use of technology within the data acquisition workflow. For example, in certain contexts, data can be collected automatically by a sensor detecting activity in realtime in its vicinity or it could be collected manually using a template available on a tablet computer. The framing helps identify the best mobility solution for data acquisition.

Lowe concluded by observing that maps are typically created for people who are already in power, who control the resources to create these documents. New technologies may make it possible to flip that equation around. For example, community mapping using volunteers can empower people in communities by identifying emergent issues, grounding conversations in context, and depicting local knowledge and values.

DISCUSSION

Kevin Brownawell, interagency professional in residence, US Institute of Peace (USIP), cited some of the difficulties that can arise in using new technological capabilities for data sharing. First, some of the most useful data that can be collected are subjective and designing survey instruments to collect this kind of information can be much more labor intensive than collecting objective data such as immunization rates. Similarly, data input can be very labor intensive. A large and well-trained staff is generally required to input large volumes of data, regardless of technological capabilities. At the same time, the quality of the data needs to be assessed, which requires an investment of time from well-trained personnel. Finally, interpretation of the data can be difficult and contentious. "Who is going to interpret the data? What type of framework do they have?" At USAID, he said, he and his colleagues often worried about passing controversial data up the chain of command, because senior officials had a tendency to interpret the data in ways that reflected their circumstances rather than the context in which the data were gathered.

Keleher agreed that subjective data are often the only data that can be collected given the focus of his organization's work. However, these qualitative data can help interpret more quantitative measures. While he agreed that methods of collecting subjective data can be methodologically rigorous, he observed that new technologies can ease data-gathering demands. For example, when a delivery man makes a delivery, it is recorded using a simple hand-held device, and the information thus collected can be valuable in an organization's decision making. Thus, a major component of an organization's planning should be careful decisions about what are the important data to collect and how to collect those data.

In response to a question about data reliability, Vinck talked about some challenges in crowdsourcing. For example, in some parts of Liberia, cell phone ownership is much lower than in other parts. "In terms of reporting, that has a major impact," he said. In such situations, there are substantial advantages to having trained people gather data, despite the greater effort required for training and sending them into the field.

He also advocated that peacebuilders learn more about how to maintain the quality of data collection and analysis. To check the data they collect, researchers can triangulate information from different sources. The technologies used to collect data also make it possible to check consistency and the reliability of interviewers and the information they gather. Research protocols have strict standards concerning how to select interviews, how to conduct them, and how to get consent. Peacebuilders also need to understand research design and the problems with flawed research approaches. "Training needs to be done on how to use and access data and also how to judge and understand data."

Richard Boly, the director of e-diplomacy at the State Department, agreed that a centralized database under the control of a single entity is not feasible, and added that citizen-generated data can both validate data generated by the government and result in data generated independently from the government. This open model of data sharing can support not only decentralized data gathering but also decentralized analysis.

Lowe emphasized the importance of the metadata description of an information asset so that it is searchable and accessible from a variety of interpretive stances. Good metadata allow data to have a much longer life cycle and greater usefulness. He also cited the importance of multiple interpretations of data, which require that the data be capable of being pulled apart and being used in a different way. This may not mesh well with current business models, but new technologies allow this kind of open-source data gathering and analysis, which create great promise for the future.

Susanna Campbell, research fellow at the Saltzman Institute for War and Peace Studies, Columbia University, noted that the uses to which data are put may constrain the selection of data collected and analyzed. She pointed to four kinds of uses of data for peacebuilding. The first is to improve the effectiveness of programming, generally through the monitoring of ongoing programs. These data are generally not shared, because people are less likely to provide full assessments if they know that what they say will be freely available. The second is to categorize peacebuilding successes and failures. These data are more likely to be shared because they are more likely to be part of an academic study than an effort to improve programming. The third is to improve coordination, which often relies on open data from the community. The fourth is to improve the targeting of programming, which requires data on the context in which programming occurs. These data can be particularly useful in demonstrating the interrelationships among systems and how systems work together.

Anne Ralte, senior advisor in the Office of the Director of Human Resources for USAID, mentioned the Standardized Monitoring and Assessment of Relief and Transitions (SMART) system, a USAID initiative.[3] As indicators for humanitarian systems, the system uses the overall mortality rate, which is a crude and somewhat controversial indicator, and the nutritional status of children under age six. Many organizations have bought into the effort, and the data are now housed in the Center for Research on the Epidemiology of Disasters in Brussels. Data are gathered by NGOs and validated by a group of independent epidemiologists, with a simple-to-use and standardized data-gathering tool. Graphical presentations of the data have been developed to improve interpretation and dissemination.

Vinck mentioned another project called Food and Nutrition Technical Assistance (FANTA), which uses standardized indicators of food security and nutrition.[4] He noted that even something as straightforward as perceptions of security measured every three years can be a very useful measure, providing a baseline against which to gauge progress (or lack of it).

Hartwell observed that maps can be highly incendiary—for example, by drawing attention to disputed boundaries. Chip Hauss, director of the Alliance for Peacebuilding, responded that maps in and of themselves are not dangerous: it is how they are used that can create disruption. Maps can be very efficient and effective tools, but, like statistics, maps can also lie.

Andrew Robertson, senior program officer at USIP, pointed out that in the workshop's morning discussions, trust was described as coming from dialogue, whereas in the current session, trust comes from method and structure. Method and structure point to the need for planning and the ability to predict relevant questions, but adaptability, flexibility, and learning are also crucial to successful peacebuilding. This shift is happening in the commercial world, from a structured to a more flexible and adaptive approach. Tools therefore need to be quickly adaptable to adapt to changes in what stakeholders think they need. This is where the morning and afternoon discussions could fit together, he said.

---

[3] More information about the system is available at www.smartindicators.org/index.html.
[4] For more information, see www.fantaproject.org.

Andrew Blum, director of learning and evaluation at USIP, observed that even domestic data-gathering projects may have lessons for peacebuilding. For example, efforts to gain information on immigrants or families with at-risk children have many commonalities with peacebuilding data-gathering efforts. Collaborations among organizations doing different kinds of data sharing and dissemination would enable the sharing of lessons learned and best practices.

At the end of the session, Roman stressed the importance of incentives for different parties to share information. New capabilities depend critically on improving the information flow among .org, .gov, and .com information domains. Thus, collaboration in a decentralized framework will be essential to the creation of data-sharing mechanisms for peacebuilding.

5

Demonstrating Data Sharing: The UNITY System

UNITY is a data-sharing, visualization, and collaboration platform developed jointly by DOD and USAID to make visible the scope and scale of humanitarian development, security assistance, and peacebuilding investment worldwide. By making key data visible to peacebuilders, UNITY can maximize allocation of scarce resources and, it is hoped, improve outcomes for the society receiving the peacebuilding intervention.

UNITY was described and demonstrated at the workshop by Mark Hainsey, project leader at the US Army Corps of Engineers, and Steven Wood, support lead for the Cooperative Security (CS) Joint Capability Technology Demonstration (JCTD) program. JCTDs are intended to exploit mature and maturing technologies to solve important military and civilian-military problems and to concurrently develop the associated concept of operation, a guidance document for the technology's users. These capabilities and operational concepts are then evaluated in exercises on a scale large enough to clearly establish operational utility and system integrity. Emphasis is on technology assessment and integration. After the presentation by Hainsey and Wood, workshop attendees offered comments on the system and ideas for its enhancement.

THE UNITY SYSTEM

UNITY is designed to provide data visibility for the US government, nongovernmental organizations (NGOs), international organizations, and other entities working in fragile environments. Organizations working in post-conflict zones tend to manage planning independently. Consequently, project activities become siloed, and interventions can become uncoordinated and inefficient. UNITY aggregates nonclassified data from partner organizations and displays it through a Web-based browser interface as a series of overlays, charts, graphs, and tables that are geospatially referenced to a map of the region. As shown in Figure 5-1, each peacebuilding project is represented by an icon that can be clicked on to see more information on budget, partners, and other project details.

UNITY was developed under the Cooperative Security JCTD, an interagency research and development initiative overseen by the Office of the Assistant Secretary of Defense for Research and Engineering. Partner agencies are USAID, the US Southern Command, the US European Command, and the US Army Corps of Engineers. The objective of this JCTD, said Hainsey, is to develop cutting-edge capabilities to produce a better integrated "whole of government" approach to development and defense cooperative activities both with other US agencies and across the public-private divide.

The problem the UNITY system addresses is the absence among regional stakeholders of integrated, interagency adaptive planning, decision support, and assessment capabilities; information-sharing architectures; and orga-

FIGURE 5-1 Hypothetical USG and NGO Development Projects in Guatemala and Belize Shown on UNITY.

nizational structures needed to conduct effective cooperative security and partner capacity-building activities. Planning is stove-piped within agencies, resulting in overlapping solutions and wasted resources. Requirements for the new system were that it use nonclassified information in non- and precrisis environments and that it engage public sector stakeholders. The system "is not a holistic solution for all of the challenges that we've discussed through the morning and early afternoon" but rather "an opportunity to start looking at tools and techniques for how to share data, how we collaborate, and how we provide mutual visibility to our partners and stakeholders," said Hainsey.

The UNITY system allows regional and multinational nonclassified information sharing, mutually visible situation and event assessment and planning, and collaborative implementation, monitoring, and evaluation. It connects communities of interest through a federated, collaborative forum called the All Partners Access Network (APAN). For the first time, users will be able to juxtapose DOD and USAID country plans to show overlaps and gaps in their planning processes. It takes in authoritative project data and categorizes it by sector based on the Department of State's Foreign Assistance Framework. And the system is scalable, with the capacity to include all federal agencies and departments that conduct foreign assistance activities as well as nongovernmental sources of information that choose to share their project data.

Platform tools enable users to access information in their areas of interest, and an innovative RSS reader discerns what individual users may want to see and custom-tailors the information provided by the system. A dashboard, tailored to a user's profile, can be configured to enable faster and more efficient visualization of information in their area of interest. Data can be sorted, filtered, searched, and displayed according to the user's preferences. The original format and syntax of the data are retained so that information from the data providers is not lost. A fine-grained data access control system allows some data to be shared only with small groups or individuals, while other data are freely available. The system is government owned and does not have proprietary components and consequently, said Hainsey, can be easily updated should new capabilities be required.

DISCUSSION

Several workshop participants applauded the capabilities and scope of the UNITY system, describing it as a valuable tool for information gathering and planning. For example, Michael Shipler, senior program advisor, Search for Common Ground, pointed out that the system provides a way to develop a shared understanding of what is happening in a country and thus could be used not just by civil society groups but by the media, local government officials, the police, and others to assess local situations and capacities.

Participants also offered feedback on the characteristics of the system, which Elmer Roman, oversight executive, Office of the Secretary of Defense, US Department of Defense, Anne Ralte, senior advisor, Office of the Director of Human Resources, US Agency for International Development, and Hainsey and Wood, who were involved in the system's development, welcomed as input for future changes. Among the suggestions proposed were software tools that could be added to the system. One useful tool would be software that could find patterns in seemingly unrelated data. For example, as Melanie Greenberg, president and CEO, Alliance for Peacebuilding, noted, if everyone in a region is suddenly buying AK-47s, access to data documenting that trend could help to prevent the outbreak of hostilities.

In response to a question about what other kinds of data can be entered into the system, Wood said that information from authoritative sources will be entered into the system. The system has been designed for precrisis environments. Roman acknowledged that including conflict indicators would be a valuable addition so that the effects of both development and peacebuilding could be monitored over time.

Participants also noted potential shortcomings in the UNITY system. As Lisa Schirch, professor, Center for Justice and Peacebuilding, Eastern Mennonite University, pointed out, much of civil society does not want the military to engage in development because of its negative impact on NGOs working on the ground. The military and other organizations may have different perceptions about the problems that need to be solved and how to solve them, and unless organizations are on the same page it can be difficult to share information. "We wouldn't want to share information if we have different goals," she said. "We do want to have conversations about those different goals and different analyses."

Another problem cited by several participants is that the tool lacks a way for countries included in the database to provide feedback. A system containing information only from US agencies seems to imply that the countries being mapped need outsiders to gather and share information. It also does not represent the full range of activities, including those of the host countries. More generally, Shipler noted, within countries different actors might identify different sorts of data as vital to their planning processes. There does not appear to be a way to manage these differences in UNITY.

Kevin Brownawell, interagency professional in residence, US Institute of Peace (USIP), observed that, though Americans may be interested in the acquisition and sharing of data, that is not necessarily the case in other countries. As a result, several basic questions need to be answered: Does everyone agree that data should be collected and shared? If so, what kind of data? With whom should data be shared? Are data open or closed? How will data be used? Brownawell suggested starting with the posting of country statistics generated by the US government and then seeing how far the system can expand into data provided by the NGO community and other countries.

Roman countered that many of the countries in which he has worked have been eager for the military to engage in development activities. The military understands its role, he said, and sharing information can allow development and stability to progress. Also, the military wants to show, in part through this program, that it can be a responsible partner in development as part of the security cooperation plan. "The more you know and understand and the more you understand what others are doing, the better it is for the unity of effort overall," he said.

Marcia Hartwell, visiting scholar, USIP, also pointed out that, although she is not a fan of military involvement in development projects, the situation varies from country to country. In some cases, NGOs prefer the military to be active. For example, in Iraq, the military secured a perimeter within which everyone working on humanitarian aid could operate safely. She also acknowledged that the military excels in dealing with other armed groups and military organizations. "Civilians work well with civilian groups, and the military works well with armed groups," Hartwell said.

In response to participants' concerns about placing their data on a platform hosted by DOD servers, Roman and Hainsey both observed that UNITY could be hosted outside the DOD environment and that both DOD and USAID have been looking for opportunities to do so. In particular, Roman suggested that USIP might be an excellent place to host such a data-sharing program.

Another potentially valuable source of data cited by workshop participants is the information available through crowdsourcing. Crowdsourcing techniques could be used to survey the viewpoint, priorities, and perceptions of the ultimate beneficiaries of development in a country—the people. However, this information, too, would most likely be accepted if available through a system not involving the military.

Patrick Vinck, research scientist, Harvard Humanitarian Initiative, noted that many platforms offering data related to peacebuilding are emerging.

Ways need to be found to enable these platforms to talk with each other, and the UNITY system could help make such cooperation possible. Other participants similarly pointed to other systems that provide complementary information, including conflict indicators. Linking to these other sources may be a better way of sharing information than having it compiled in a system developed by the military.

Vinck also asked whether UNITY or some other collaborative software system could be made more open and sharable. Wood explained that the platform will support multiple server configurations, so extensibility has already been built into it and multiple versions of the platform could be spread across multiple user communities.

Appendix A

Agenda

Workshop on Using Data Sharing to Improve Coordination in Peacebuilding
of the
National Academies and
United States Institute of Peace

May 23, 2012

U.S. Institute of Peace
2301 Constitution Avenue, NW
Washington, D.C.

The goals of this workshop are to identify what needs a data-sharing system must address in order to secure the participation of other federal agencies and nonfederal organizations with a role in peacebuilding, and create more effective coordination in conflict zones.

8:00 a.m.   **Breakfast**

8:30 a.m.   **Welcome and Goals for the Day**
Melanie Greenberg, Alliance for Peacebuilding; Elmer Roman, Department of Defense

9:00 a.m.   **Defining the Challenges of Coordination**
Despite rapid change in the capacity of technologies to manage and analyze data, coordination among government and nongovernment actors in peacebuilding interventions remains a challenge. Why? In this session, we will discuss the principal technological, ethical, and cultural challenges faced by peacebuilding actors as they struggle to break down the organizational information silos in which they work.

	Speakers:	Robert Loftis, USIP
		Randy Tift, World Vision International
		Susanna Campbell, Columbia University
	Moderator:	Melanie Greenberg, Alliance for Peacebuilding

10:15 a.m.   **Break**

10:45 a.m.   **Overcoming the Challenges to Sharing Information**
Across multiple crises and interventions in Iraq, Afghanistan, and Haiti, the USG and NGO communities have improved processes for managing inter-community interaction by developing guidelines to define when and where cooperation is possible. These guidelines enable cooperation while protecting both groups' independence and security. What are the characteristics of a data-sharing process that would enable cooperative planning while securely maintaining executional autonomy?

	Speakers:	Lisa Schirch, Eastern Mennonite University
		Marcia Hartwell, USIP
	Moderator:	Melanie Greenberg, Alliance for Peacebuilding

12:15 p.m.   **Lunch**

1:00 p.m.   **Using Data for Impact**
Establishing the processes that allow collection and sharing of data between organizations can be costly. Furthermore, not every stakeholder in the sharing process may see the same value. Monitoring and evaluation is a crucial element of the project management process that can increase the benefits from data sharing. Can innovations in data acquisition, data analysis, and data visualization when used properly create value that positively incents broader participation in the data-sharing process?

	Speakers:	Stephen Lowe, USDA
		Niall Keleher, Innovations for Poverty Action
		Patrick Vinck, Harvard Humanitarian Initiative
	Moderator:	Elmer Roman, Department of Defense

APPENDIX A

2:30 p.m. **Demonstrating Data Sharing – The UNITY Platform**
In partnership, the Department of Defense and USAID have developed a platform for sharing unclassified planning information. By making the scope and scale of humanitarian and peacebuilding investment in a given conflict zone visible, the goal is to enable better allocation of scarce resources and thereby improve outcomes in the society receiving those resources. This session will demonstrate how even limited information sharing can positively affect intervention outcomes.
Speakers: Mark Hainsey, USACE
*   Steven Wood, Integrasure*
Moderator: Elmer Roman, Department of Defense

4:00 p.m. **Final Thoughts**
Melanie Greenberg, Alliance for Peacebuilding; Elmer Roman, Department of Defense

4:30 p.m. **Adjourn**

Appendix B

Attendees

Roundtable Cochairs

Melanie Greenberg
President and CEO
Alliance for Peacebuilding

Elmer Roman
Oversight Executive
Office of the Secretary of Defense
US Department of Defense

Steering Committee Members

Gregor Bailar
Chief Information Officer (ret.)
Capital One Financial Corporation

Mark Hainsey
Project Leader
US Army Corps of Engineers

Chip Hauss
Director
Alliance for Peacebuilding

Suzanne Kindervatter
Vice President
InterAction

Stephen Lowe
Geospatial Information Officer
US Department of Agriculture

Phuong Pham (not attending)
Research Scientist
Harvard Humanitarian Initiative

Michael Shipler
Senior Program Advisor
Search for Common Ground

Speakers and Expert Participants

Richard Boly
Director of e-Diplomacy
US Department of State

Willie Brandt
Division Chief, Stability Ops and Civil Support
Joint Chiefs of Staff

Robyn Broughton
Program Economist
Office of Afghanistan and Pakistan Affairs
US Agency for International Development

Kevin Brownawell
Interagency Professional in Residence
US Institute of Peace

Susanna Campbell
Research Fellow
Saltzman Institute for War and Peace Studies
Columbia University

Aaron Chassy
Senior Technical Advisor
Governance and Civil Society
Catholic Relief Services

Gerard Christman
Program Manager
Office of the Secretary of Defense
US Department of Defense

Jim Garcia
Research Analyst
Center for Technology and National Security Policy
National Defense University

Eric Gundersen
President and Cofounder
Development Seed

Marcia Hartwell
Visiting Scholar
US Institute of Peace

John Holloway
HADR Communications and Information Sharing Analyst
Office of the Chief Information Officer
Department of Defense

David Kamien (via phone)
Founder and CEO
Mind Alliance

Niall Keleher
Director of Research Methods and Training
Innovations for Poverty Action

Robert Loftis
Interagency Professional in Residence
US Institute of Peace

Julie Montgomery
Director
Innovation and Learning
InterAction

APPENDIX B

Anne Ralte
Senior Advisor
Office of the Director of Human Resources
US Agency for International Development

Lisa Schirch
Professor
Center for Justice and Peacebuilding
Eastern Mennonite University

Matt Scott
Director
Peacebuilding
World Vision International

Shannon Scribner
Humanitarian Policy Manager
Oxfam America

Randall Tift
Senior Policy Advisor
World Vision

Jessica Vogel
Director of Programs and Operations
International Stability Operations Association

Linton Wells
Director
Center for Technology and National Security Policy
National Defense University

Patrick Vinck
Research Scientist
Harvard Humanitarian Initiative

Staff

Geneve Bergeron
Research Assistant
US Institute of Peace

Andrew Blum
Director of Learning and Evaluation
US Institute of Peace

Sheldon Himelfarb
Director, Center of Innovation for Science, Technology and Peacebuilding
US Institute of Peace

Steve Olson
Freelance Writer

Greg Pearson
Senior Program Officer
National Academy of Engineering

Proctor Reid
Director of Programs
National Academy of Engineering

Andrew Robertson
Senior Program Officer
US Institute of Peace

Frederick S. Tipson
Special Advisor
Center of Innovation for Science,
    Technology and Peacebuilding
US Institute of Peace

Audrey Warren
Administrative Assistant
Rule of Law Center
US Institute of Peace

Steven Wood
CS JCTD TM Support Lead
Integrasure/US Army Corps of
    Engineers (support contractor)